A PLUME BOO

A FIELD GUIDE TO HOUSEHOLD BUGS

JOSHUA ABARBANEL and JEFF SWIMMER are Double J Media, based in Santa Monica, California.

Swimmer is a journalist and a writer/producer of documentary films for PBS, BBC, National Geographic, and Discovery. His films cover a range of topics in natural history and popular culture.

Abarbanel uses a variety of digital tools and software to create large-scale works on paper that have been exhibited in Los Angeles and New York. In addition, he designs fabric patterns for the textile industry, and teaches a range of digital arts and media classes at Los Angeles Harbor College. He lives in Santa Monica with his wife and two children.

A Field Guide to Household Bugs

IT'S A JUNGLE IN HERE

Joshua Abarbanel

Jeff Swimmer

A PLUME BOOK

PLUME
Published by Penguin Group
Penguin Group (USA) Inc., 375 Hudson Street, New York, New York 10014, U.S.A.
Penguin Group (Canada), 90 Eglinton Avenue East, Suite 700, Toronto, Ontario, Canada M4P2Y3
(a division of Pearson Penguin Canada Inc.)
Penguin Books Ltd., 80 Strand, London WC2R 0RL, England
Penguin Ireland, 25 St. Stephen's Green, Dublin 2, Ireland (a division of Penguin Books, Ltd.)
Penguin Group (Australia), 250 Camberwell Road, Camberwell Victoria 3124, Australia
(a division of Pearson Australia Group Pty. Ltd.)
Penguin Books India Pvt. Ltd., 11 Community Centre, Panchsheel Park, New Delhi – 110017, India
Penguin Books (NZ), 67 Apollo Drive, Rosedale, North Shore 0745, Auckland, New Zealand
(a division of Pearson, New Zealand Ltd.)
Penguin Books (South Aftrica) (Pty.) Ltd., 24 Sturdee Avenue Rosebank,
Johannesburg 2196, South Africa

Penguin Books Ltd., Registered Offices: 80 Strand, London WC2R0RL, England

First published by Plume, a member of Penguin Group (USA) Inc.

First Printing, October 2007
1 3 5 7 9 10 8 6 4 2

Ⓟ REGISTERED TRADEMARK—MARCA REGISTRADA

CIP data is available
ISBN 978-0-452-28874-4

Printed in the United States of America
Set in Optima
Designed by Joshua Abarbanel and Jeff Swimmer, with Helene Berinsky

To our love bugs Gayle and Stacey

And all of our little bugs:
        Ella, Dylan, Colin, and Juliet
        Aliza and Chloe

CONTENTS

ACKNOWLEDGMENTS

We would like to thank the many relatives, friends, bug experts, and veteran writers who helped us, inspired us, and enabled us to create this book.

Much of our inspiration to create this book came from the wondrous work of the many talented electron microscope photographers who bring us so shockingly close to the bugs and their amazing worlds, and whose images make this book feel so alive.

Thanks to the patient librarians at the Santa Monica Public Library; Marty Katz, Craig Cooper, and Lou Weitzman, who helped us sort out the legal fine print; Paul Schneider, a fellow writer who urged us onward; and Lisa Cooper, who overcame her squeamishness about the topic to help us with research. A big thanks to entomologist Steven Kutcher, who fact-checked this book for us.

We are most grateful to Megan Newman and Trena Keating for helping us get this project off the ground. Warm thanks to Amy Hill and her graphics team, and of course our most helpful and patient editor, Hilary Redmon.

Last, a deep thanks to the many friends and family members who gave us the confidence and strength to keep at it. Though they often winced at the images and grisly facts in the book (as did we!), they heard us out when we promised them we might be on to something, and then had the audacity to agree.

Jeff & Josh

INTRODUCTION

We love bugs as much as the next guy. They remind us of the infinite variety and majesty of nature: flattened on the windshield, spread-eagle on the sole of our shoe, dangling from the end of a newspaper. We just love 'em. We'd like them sautéed too, just as long as they're not on our plate.

So when Penguin asked us to write a book about bugs (OK . . . we asked *them* if we could), we knew it would be a stretch. We took a lot of classes in college, some of which we even remember—but nary an entomology class between us. For UC Berkeley grads like us, we weren't gonna study them unless they were wearing Che Guevara T-shirts around their antennae.

What really got our bug juices flowing is the paradox of that iconic American image called home. To most people, home has always been a symbol of calm; it's where the proverbial heart is, and it's a refuge from a dangerous world "out there." But what if home really isn't such a sanctuary after all? (And here's where the bugs come in.) What if nature, in all her beastly, hairy, tentacled, buggy glory really rules at home— and we just crash there?

One hint that this could be true comes from no less a reliable source than the United States Food and Drug Administration (FDA). In its deliciously euphemistic "Food Defect Action Levels" report, it assures the public that all of the following food "impurities" are harmless and require no regulatory action:

60 insect fragments or 1 rodent hair per 100 grams of CHOCOLATE, 1 rodent excreta pellet per unit of POPCORN,

13 insect heads per 100 grams of FIG PASTE, 1 milligram of mammalian excreta per pound of PEPPER, 10 fly eggs per 500 grams of CANNED TOMATOES . . .

And on and on. To read the FDA report is to laugh, cry, and dry heave at the same time.

So this nagging possibility—that we're just guests in our homes—led us to want to meet the "real" occupants of our homes, which led to this book. When poring over the stunning electron microscope photographs of the bugs you see in these pages, we were amazed at the variety of expressions, body armor, odd dangling appendages, mega-eyeballs, and sinister mouthparts. They seemed both shockingly human and monstrous at the same time. And to know they live in our pillows and sheets, our eyelashes, our couches, under our floorboards, and inside our pantries (and sometimes panties!)—well, that's just more than we could bear. It was easier to off-load all this awful knowledge into this book, than to keep it all inside of us. So, apologies in advance and thanks for taking the load off.

When it came to the bug world, we journeyed from fear to respect, from shock to awe, and, always goggle-eyed wonder. We've gone out of our way to stay away from the "Ripley's Believe It or Not" approach of focusing on the most bizarre bug behaviors on Earth that almost no one will ever actually see or deal with. This book is the opposite—a practical field guide to our most intimate home companions, the day-to-day, night-to-night critters with whom we share this place called home.

We hope you'll have as much fun reading it as we had writing it.

Jeff & Josh

A Field Guide to Household Bugs

Bedbugs

CIMEX LECTULARIUS

The frisky, bloodsucking *Cimex lectularius* has been around for at least 3,500 years, which means that the ancestors of bedbugs snacking on us today may have feasted on the royal blood of the pharaohs, or even Caesar himself.

This scrappy parasite will shove his long, skinny proboscis into his host (um . . . you), and suck away until he's had enough. After walking it off with a little stroll on your skin, he'll shove his beak in again for a nightcap, taking three or four more good belts before hanging it up for the night. And with a schnozz one hundred times finer than a hypodermic needle, you won't feel a thing.

But it's not just fun and run for this nocturnal naughty boy. He'll leave his business card: three tiny blood-caked parallel lines (called "breakfast, lunch, and dinner" in the trade), up and down your arms or legs.

Don't worry, though; these guys may turn your chest into Dracula's wet bar, but they don't carry any diseases that transmit to humans.

APPEARANCE

A wingless, one-quarter-inch-long, flat, and oval-shaped body. Six legs, large antennae, and big mandibles. Bedbugs, like many of us, have a hard time keeping their weight steady. After a good snack, they turn rust red and balloon to about the size of this *O*, though their svelte frame at birth is barely bigger than the period at the end of this sentence.

DIET

Blood is all they eat. They can live for a year without a meal, and if humans aren't around, they can feast on cats, dogs, chickens, birds, mice, rats, rabbits, and guinea pigs. They sleep all day in mattresses, wall cracks, and the like, and wake up at night while you sleep. A bedbug must go through five molts to advance to each of its next life stages. Each requires a blood meal first. So, in a way, their feasting is a kind of "blood mitzvah."

SMELL

They survive by giving off ultrasmelly pheromones from a pouch near their hind legs. They use the smell to find mates and warn others of danger. If there are enough in your house, the pheromones create a sweet, musty odor that some have compared to the smell of fermenting raspberries. French people call them *punaise* ("stinking") for their stinky ways. And it's not clear just why they do. Their smell appears to play no role in any aspect of mating. In fact, scientists have found that revved-up males will try to mate with a piece of cork carved in the shape of a bug.

MATING

When they do mate, the act is so violent that entomologists refer to it as "traumatic insemination." Females have no genital opening, so a male will force himself into her abdomen, cutting it open to deposit sperm. The force is so strong that females have developed a soft tissue mass to absorb the blows. The sperm then finds its way to the reproductive organs.

REPRODUCTION

Females lay up to five hundred eggs in a lifetime, or a few a day. The eggs are coated with an ultrasticky cement that clings to any surface. Even after the bug has hatched, the eggshells remain cemented in place. Hatchlings require blood feeding before they can reach adulthood.

THEY GET AROUND

Bedbugs love joyriding—on your coat or shoes, as you make your way to work, your friends' houses or beds, maybe sometimes to the in-laws. And, at every stop, they may decide to park it for a while and spread the love.

GET THEM OUTTA HERE!

At one point, people used "assassin bugs" (subfamily Triatominae) to kill off bedbugs. It worked really well, until the bedbugs died out and the assassin bugs started attacking humans (and hurting them far worse than bedbugs ever did).

They like clean homes, so forget about cleaning these bug-gers away. And store-bought pesticides? Yeah right . . . it's chum to these guys. And these bugs do snug, too—they can live happily in a space no wider than a playing card.

The classical Greek philosopher Democritus suggested hang-ing a dead stag at the foot of the bed to keep the bugs away. Later, bedbugs may have spread across Europe and the Unit-ed Kingdom with the legions of Julius Caesar. Now they're worldwide.

The word *bug* has several origins, among them a Middle Ages word for bedbugs. *Bug* was a Celtic word for a ghost or goblin, and bedbugs were considered terrors of the night. In Egyptian villages they're called *akalan* ("an itching") and in Sanskrit *uddamsa* ("biter").

One famous entomologist who wanted to collect bedbugs for research would stay in motels, set his alarm for 2 A.M., and simply wake up and collect all the samples he wanted from his own bed.

An 1870s pest control guide recommended rubbing bedframes with spirits of turpentine and kerosene oil, and filling the cracks in floors and walls with hard soap. The guide also advised putting a quarter pound of brimstone (sulfur) on a dish in the middle of the room and lighting it. This would bleach the walls, thus kill the bugs. One bedbug remedy popular in the 1930s called for using a chemical that was so harsh it peeled the finish off metal beds.

At the turn of the century, a treatment for various ailments called for a brew of sorghum juice, black beans, garlic, rum, and seven freshly killed bedbugs.

A modern suggestion: petroleum jelly all around the legs of the bed will keep them from crawling up to get you. Or put each leg in a container of water. Of course, this won't prevent the occasional *Mission Impossible* move; bedbugs have been known to crawl up the ceiling and let go just above the bed. Alternately, you could try to get rid of all the cracks in a room by caulking, replastering, or painting—or crank up the temperature in your bedroom to 113 degrees or higher. This may kill a few.

BY
HOOK . . .
Bedbug bite or spider bite? Spiders leave two bite holes close together, and bedbugs often leave three. Curiously, it's a salivary secretion bedbugs excrete before biting that causes the itching, not the biting itself.

The very worst bites may cause nervous system reactions, digestive disorders, and loss of sleep. Shoulders and arms are their favorite kill zones.

Bedbug claw

Oddly, perhaps a little rudely, bedbugs are happy to suck your blood straight out of your skin; but while they're sucking they don't like to touch your skin, perhaps for fear of rousing you. They prefer to hold on to your clothes or bedding. And they don't even locate you by sight—just by the warmth you give off and the carbon dioxide you exhale.

. . . OR BY CROOK

Around the world, bedbug infestations are back in a big way—even with some bugs, such as cockroaches and ants, on the retreat. Suspected culprits: increased immigration, cheap global travel, and the banning of powerful pesticides like DDT. Just twenty-five years ago they were nearly extinct in most developed countries, so why they've come back is a mystery.

In 2006, a Chicago woman (and her publicist) sued a Catskills hotel for $20 million, holding up for the press photos of herself covered in bloody bites and pustules on her back, chest, arms, and legs. She claimed that after three nights of bedbug mauling, she felt as if her skin "was on fire" and wanted to "tear it off" (CNN.com, March 7, 2006).

One crazed victim in New York City switched to white sheets so she could see the bedbugs better, and set up a bedbug "jail" in a Tupperware container that she put out on her windowsill to "torture them with daylight" (*The New York Times,* November 27, 2005).

DELUSIONAL PARASITOSIS

Just because you think you are being bitten doesn't mean you are. If this sums up your state of mind, to an extreme you may be suffering delusional parasitosis. Please seek help.

Lice

CAPITIS HUMANIS

Lice are a bit like Rod Stewart. They love hair, and they keep us up at night in a twist. But similarities stop there. "Lousy" heads make us feel exactly so, and these parasites are so beautifully designed to feast on hair and necks—especially on kids—that school nurses will probably never see the end of them.

Kids get them, schools fret them, and parents fight them with an arsenal of soaps and home remedies. But lice have brilliant ways of hanging around. Their six strong legs and special claws are ingeniously designed to cling to hair shafts.

Kids are perfect hosts. As kids tumble and play together, the lice spend lots of time leaping back and forth from head to head. But take heart: blood transfusions won't be necessary. Lice will only drink about $1/10{,}000$ of a milliliter of Junior's blood at a meal and few kids ever harbor more than ten at a time. (For more precise figures, Harvard's School of Public Health, has a lice blood loss calculator.)

Head louse

As a child-focused pest, it's not surprising that kids have taken up the word *cooties*, a lice nickname, to describe playground rivals' unsavory traits.

It may be some comfort to know that lice can't jump or fly or burrow into the scalp. And they rarely pass on major infections. But that's not much consolation. These gray-coated beasts have a license to bug, causing terrible itching and sleeplessness. In addition to the bites, their spit and poop can cause itching, not to mention the psychological factor of knowing they are there.

APPEARANCE
With their menacing features and body armor, they look like alien predators seen through the wrong end of a telescope. Just knowing they're coming after the brood is enough to bring on the sweats.

DIET
Lice require at least one blood meal a day, and can't survive even a day at room temperature without one. So if you can keep them at bay for a day, you could conceivably kill them, and then pick them and their eggs off one by one. But this would, of course, make you a nitpicker. (Nit = louse egg.)

REPRODUCTION

Some females will lay fifty to three hundred eggs in their two-week lifetimes, averaging about five a day. Since their body temperature is constant, they can hatch eggs any time of day or year.

Cattle louse

LIFE CYCLE

After an eight-day gestation, a baby louse will hatch and start feeding right away. It needs a blood meal within minutes of birth to survive.

After nine to twelve days, it's a full adult. It will eat, raise holy hell, have lots of babies, and then move to Florida to play golf and keel over in about two weeks.

HOW THEY SPREAD

Best spread by proximity, or shared combs or hats, they can also live on the bedding of a human host. Most spend their entire lives in our hair.

CRAZY COUSINS

The head louse's cousins include pubic lice, which live just where you'd expect them to. Another cousin is the body louse, which is less common but has a useful advantage: unlike the head louse, it can live on clothing for days, and can comfortably traverse bald parts of the body in search of hairy havens. Slobs beware: body lice may take root on those who don't do their laundry often enough! Body lice are potentially much more dangerous than head lice, as they can carry typhus, trench fever, and other—sometimes fatal—infections.

THE THINGS THEY CARRY

Pediculosis is the word for "infestation with lice." Just a dozen or so lice on your head constitutes an infestation, though some sufferers are likely to host hundreds of dead ones on their heads as well.

GET THEM OUTTA HERE!

Lice are tenacious. They've been found on prehistoric mummies, and have had lots of time to adapt to living on us. So it's no wonder they're so good at hanging around. Oddly, they seem to like Caucasians more than any other ethnic group, possibly due to Caucasians having thinner hair.

Nit on human hair

Treatment is a horrid mess, involving individually plucking out the offenders from the hair with tweezers or a similar device.

And it's easy to tweeze at the wrong offender—like dandruff or hairspray droplets and even dead lice eggs. Remember, lice move.

Combing out lice is also useful. It must be done every day for two weeks—with a magnifying glass and very bright lights—until no more lice are visible.

Head louse

Head louse on human hair

Electronic louse "zappers" are trendy,
but in the end, no more useful than combs.

If you can't tweeze 'em, oil 'em. Some say olive oil or hair
gel can suffocate them, but drowning one's hair in such
stuff can cause new problems. Ordinary shampoos only
make your lice clean. You could also try baking them to
death with a jet of air from a hair-dryer, or if you're desper-
ate you can shave off your kid's hair. This will work.

The liquid poisons pyrethrin or permethrin might also
work, killing the lice ten to thirty minutes later.

Paranoids only: lice have been known to lay low in kids' car seats, so vacuum away. And, of course, clean their sheets, clothes, stuffed animals, bike helmets, shared headphones, etc.

Dust Mites

DERMATOPHAGOIDES PTERONYSSINUS

When you lay your head down on your pillow, you're never alone. You're sharing your mattress with anywhere from 100,000 to 10 million eight-legged pals. In fact, 10 percent of the weight of a two-year-old pillow may be composed of dead dust mites and their droppings.

But don't worry if you leave bread crumbs in bed. They won't touch those. Mites (*Dermatophagoides* = "skin eater") prefer feasting on your dead skin. Even though they have no eyes, they can home in on you from many feet away. And you ought to thank them for that. If they didn't eat your skin, it would pile up all around you. (We all shed about a gram of dead skin every day.)

Are you allergy prone? Your bedmates might have something to do with that. These puppies see more costume changes than a Madonna concert. Each time they throw off their own dead skin, they sprinkle it across your bed and pillow. Do you enjoy cuddling up in bed with a dog or cat? Yummy, pass the salt. Mites love pet skin almost as much as they like yours.

Think you can whisk them away with regular cleaning? Not so fast, sister. Each time you whip those sheets through the air, you send them—and their skins—on a joyride, spreading them throughout the room.

APPEARANCE

No, that's not a wardrobe malfunction. Dust mites have translucent, or see-through bodies. You'd probably hate them a lot worse if you could see them. Luckily, they're just $1/100$ of an inch long.

DISEASES

One reason house dust is such a common allergen is that it is so heavily infested with mite poop and mite skins. The digestive juices gurgling out of mite guts and dust mite allergens are thought to be a factor in 50 to 80 percent of cases of asthma, hay fever, and eczema.

FAVORITES AND PET PEEVES

Favorite snack: dandruff. But only dandruff with less than 10 percent fat. Luckily, there are fungi that predigest the dandruff so the fat content is low enough for the dust mite to eat. Of course, the fungus soon becomes a competitor with the dust mite for food—a contest to make Darwin proud.

Favorite cafeteria: warm, humid areas filled with dust—pillows, mattresses, carpets, and furniture.

Dust mites are arachnids, not insects, so they have eight legs instead of six. Their closest relatives are the ear mites of cats and dogs, and some sheep mites.

Creepy fact: dust mites' favorite temperature is 71 degrees— the same as for humans.

GET THEM OUTTA HERE!

The mighty mite is so tough, one leading pest control company admits eradicating them is "not realistic." There are no pesticides approved for controlling them. And some cures, like tannic acid, can be worse than the mites themselves.

If these bedfellows really do keep you up at night,
pest controllers advise you to cover pillows
in polyurethane, and soak your sheets
in water heated to 140 degrees
Fahrenheit. Every week.
Oh, and avoid contact
with mites.

Eyelash Mites and Scabies

DEMODEX FOLLICULORUM AND SARCOPTES SCABIEI

Of all the beasties that feast on our flesh, there may be none more sci-fi, drag-down nasty than the eyelash mite. Most of us have up to about twenty-five of them on our lashes at any given time, and makeup users and others who apply oily substances around their eyes are most at risk.

Fat, fluttery eyelashes are a cafeteria to these yellowish, half-millimeter-long cigar-shaped curios who bury their heads (tails?) into the gap between the eyelash follicle and the lash itself. There they gorge themselves on young shoots of eyelashes and nearby cells, using the proteins in such an efficient manner that they produce no waste. Consequently, they have no rear end (take that, J. Lo!).

Do you ever feel like something is really getting under your skin? Well, you might be—literally—dead on. *Scabies* is Latin for "itch," and these soft, eight-legged critters cause bad itches—in any language. Millions of cases of scabies infestations are reported every year. Male and female scabies mites play on your skin like a submarine at sea, sometimes crawling around on the outside and—when the

Eyelash mite

mood strikes—burrowing inside to lay eggs and to feed. Dive-bombing in and out of your skin produces the worst itching, and the eggs themselves produce an intense allergic reaction as well. In one of nature's cruel double whammies, scratching those itches just causes more itching.

Our fans always ask us why we batch these two beasties together. We do this because between them—eyelash mites on the face and scabies everywhere else—they feast on our entire bodies. And we like this sort of thing.

Eyelash mites emerging from a follicle.

APPEARANCE
Scabies are eyeless, oval-shaped, covered in long hairs and spines (females), and are tiny at just under half a millimeter. And they have chunky legs. If eyelash mites weren't eyelash mites, they'd be cigarettes—stubbed out at both ends. You'd see them being smoked at Monaco roulette tables. Mighty, frighty

scabies, on the other hand, are fear itself—evil *con carne.*

LIFE CYCLE

Scabies mating happens right out there on the waterbed of your skin. After mating, the female will use the suckers on her legs, and her sharp legs and jaws, to shimmy herself under the top layer of your skin. She'll then lay two to four eggs a day for her one- to two-month

Scabies mite

life span. Scabies babies hatch in about a week, live for another month, and pass from person to person by skin contact (use your imagination). Luckily, they don't jump from person to person.

Female eyelash mites lay up to twenty-five eggs in a single hair follicle. When they mature after a few days, they wander around your face to find a partner, mate, and then look for a juicy follicle in which to lay their eggs.

Eyelashes with
mites at the base

LIFE EXPECTANCY
Sexual maturity is reached
after ten to fourteen days, just as
midlife approaches.

HABITAT
Scabies rashes or "burrows," which look like planting rows, pop up most often around elbows, wrists, armpits, and buttocks. If it's any consolation, *Sarcoptes scabiei* don't seem to like faces (perhaps because they might have to compete with eyelash mites for food!). Though they're microscopic, you might find a tiny pinprick of white chillin' in a burrow on your body. Bingo!

REALLY CREEPY
Scabies incubate on you before any symptoms occur, so you could be infecting countless loved ones during that time before you—or they—know it. This also makes eradication difficult, because by the time you know you're infected, the culprits are probably long gone.

Eyelash mites have six legs when born, and then, just as with many mites, they grow two more after a few days.

FEEDING

Both male and female
scabies wake at night to feed.
If you feel one walking around on
you (pity!), it's more likely to be a girl,
which is twice as long as a boy.

Eyelash mites (sometimes also called "face mites") have specially adapted tiny claws and pointy mouthparts for eating follicle cells and dead skin. Scales help anchor them to the hair follicle.

Eyelash mites can also mess with your tear production. A tearful thought if ever there was one.

THE THINGS THEY CARRY

Scabies on dogs are called "mange" (as in "mangy mutt"). If dog mites find their way to human skin, they'll do a lot of itch damage before dying young. Humidity and tight clothing also attract scabies.

Dogs also boast their own species of eyelash mites called *Demodex canis*, but it's a species that won't leap across to *Homo sapiens*.

If your immune system is down for any reason, you may be in line for a pernicious flavor of scabies called "Norwegian scabies," which leaves crusty scabs all over the skin, and serious inflammation.

GET THEM OUTTA HERE!

Scabies mites can't live long off their host (you). After a few days, they dry up and die. So if you can get them off you (that's next), then you can smile knowing their hours are numbered.

Eradicating scabies is all about finding the telltale zigzag line of the burrow. Once you find it, scabies can be treated with a host of topical lotions, most of them doctor-prescribed. Since it's rare for scabies to attack only one member of a family, everyone in a scabies household should be treated at the same time—head to toe and everywhere in between.

Bodies first, and then all clothing and bedding should be scrubbed in scalding water. But even if all of them are killed, dead mites linger in your skin for about another month and only disappear when your body naturally sheds its skin.

Sometimes a bunch of eyelash mites will attack a single lash at the same time, and this is when risk of infection is highest. This can bring on a condition called "demodicosis." If irritation and inflammation get really bad, it could lead to a condition called "demodex blepharitis," which can make your eyelashes fall out.

EXTRA BONUS POINTS

Try plucking out one of your own eyelashes and then looking at it under a microscope. Chances are you'll find a happy batch of mites, bottoms up in your follicles!

Scabies on human skin

Silverfish and Earwigs

LEPISMA SACCHARINA AND FORFICULA AURICULARIA

Entomologists might not agree with us lumping these two beasties together, but the more we got to know them, the more appropriate it seemed. They both love paper and starchy, gluey snacks like book bindings or wallpaper paste. These nocturnal creepers both do some sneaky side-to-side locomotion that would make Baryshnikov proud.

And they each boast some nifty backsides—silverfish flaunt three very long, skinny antennae, and earwigs a beefy hook that makes them look like a small, brown trailer hitch.

Silverfish are a constant home companion because they'll eat anything we'll eat and will chow on dead animals (or dead people). They'll even eat their own molted skin. They're unusually long-lived (up to seven years!), and females lay big broods—about one hundred eggs at once. They're flat, scaly, and shimmer with a silvery, metallic sheen. Their fishy shape gives them their name.

Though silverfish get around fine on their six legs, they prefer to be carried by people, who trot them around in boxes, books, and papers. Like many primitive insects, they have no wings. And, in a macabre game of peekaboo, they leave their discarded shells all over their stomping grounds like decoys, which fool would-be exterminators.

Silverfish

Earwigs are a cartoon of insect terror—flat and feisty, shiny, curious, and—unique in nature—hooked in the trunk. But despite their comically fearsome look, they are, in fact, harmless. In Old English *earwicga* means "ear creature." As one of our favorite insect guides points out, though, they "rarely jump in your ear." Hardly ever. Seriously.

These nocturnal nudniks get their name from a European superstition about how they supposedly enter the ears of sleeping people, tear through the eardrum, and lay eggs in the brain. Lucky for us, there's no proof this has ever happened (except in the second *Star Trek* movie), but the cute name stuck.

There was the story of a man who, in 1961, had a major earwig bite inside his shoe that caused bad bleeding. Ouch. For their occasional pinching, they are sometimes called "pincher bugs," and can even shoot out a smelly ooze when threatened.

Fortunately, an earwig's favorite snack is not us. They'll eat live or decaying insects or vegetation first, but human skin and blood will do in a pinch.

And, curiously, though only a lucky few have ever seen this feat, earwigs can fly! When it's sunny and warm out, and the mood strikes, an earwig can fly a distance of up to thirty feet.

APPEARANCE
Most common house species of earwig are the European variety, the striped or riparian earwig (¾ inch long and really stanky), and the ring-legged earwig (⅝ inch long and not so stanky). The male's pincers are much more curved than the female's. Earwigs have a cockroach-like appearance, but are not related to roaches.

BIRTH
Silverfish eggs take about a month to hatch, depending on temperature and the availability of food around them.

HABITAT
Silverfish can live wherever it's cool and damp (think laundry room, rumpus room)—in walls, under floorboards, in the dark, on a boat, in a train . . .

The only surfaces they can't master are smooth, vertical ones, like in a washbasin. They hide so well that if you see one, there's likely to be hundreds more lurking within inches. Hopelessly lazy, both of these critters will sleep all day long, just about anywhere.

Male earwigs

There are about 1,800 species of earwigs, and they can be found in ecosystems as varied as alpine snowfields and marine shorelines.

FAVORITES AND PET PEEVES

Silverfish are a librarian's nightmare. Like termites, they have enzymes that help them digest cellulose, and they love paper. Oddly, they hate dust. Favorite snack? Rayon.

But they can go for a month without eating a thing.

REPRODUCTION

The male silverfish produces a packet of sperm called a "spermatophore" when there's a nearby female who catches his fancy. She then brings it into her genital opening.

For earwigs, it isn't much prettier. Mating is "end to end," i.e., forcep to forcep. Females of some species will choose the male with the biggest forceps. And usually after mating, males will skedaddle immediately.

FAVORITE BOOK
Older classics, anything mildewed.

GROWING UP EARWIG
Unlike most insects, earwig moms care for their babies after they hatch them, by guarding their nests with her huge forceps and bringing in food or regurgitating her own to feed them. They even tend their eggs, licking them, turning them

Silverfish

over as if on a rotisserie to prevent them from getting fungal infections. But if she's in a bad mood, she won't hesitate to devour her own eggs. She's happy to rear them through the first few life stages (called "instars"), but is peeved by those who overstay their welcome—she may eat them.

NICKNAMES

Silverfish are sometimes tagged "fishmoths" or "bristletails." And, some like it hot—really hot. One silverfish cousin (of more than 370 species) known as a "firebrat" loves ovens, fireplaces, and hot-water pipes. They've probably gotten smacked on the head by rolling pins more than any non-Stooge.

WEIRD FACT

Some silverfish will raid ant and termite nests to eat their babies.

Earwigs are real neat freaks, and will lick themselves clean all day, using tongues that can reach around their entire bodies.

TAKE NO PRISONERS

Earwigs have a real weakness for rolled-up newspapers. But even though you'll be tempted, don't smear them around after clocking them. This will spread their odor.

For the faint of heart, roll up a newspaper, soak it in water, and leave it outside for a few days. When earwigs collect in the paper, shake them into the neighbor's yard. Voilà! Humane pest control.

Silverfish-stopping techniques: use toxic dust bombs, boric acid, or foggers, or plug, putty, and block any areas around pipes in a house; keep spaces clean and dry; rotate books around in a bookcase; keep food in tightly sealed containers . . . ah hell, just move out.

And they'll never tell you this, so we will: chickens love to snack on earwigs.

If you can't keep earwigs out of the house, maybe try keeping them in. They can't breed indoors (except in potted plants) and they'll eventually die off.

Silverfish

Flies

MUSCA DOMESTICA AND DROSOPHILA MELANOGASTER

Is it just us or does your kitchen magically sprout fruit flies as if from nowhere as soon as a ripe fruit is set out on the counter? We thought so. As they hover around your bananas or peaches, they almost seem to be asking, "You almost done with that?"

This phenomenon of fruit fly armies emerging as if from thin air has been called "spontaneous generation." But this is bunk. These red-eyed beasties actually do come from somewhere—but you might rather not know. And you don't ever want to play drinking games with these fruit fly critters. For their body weight, they drink fruit alcohol profusely. Talk about a cheap date: fruit flies can even catch a buzz on alcohol fumes!

Next time you get sick, there's a decent chance you picked something up from the housefly. The housefly has a leg in spreading more than one hundred disease-causing pathogens in humans, including chol-era, tuberculosis, anthrax, and a gardenful of parasitic worms. Flies pick them up from their favorite "restaurants"—garbage cans, poop piles, dead animals, and rotting food —and then carry them to you and your home via their mouths, vomit, and contaminated body parts. Thanks for spreading the love!

Housefly

APPEARANCE

Fruit flies: yellow bodies, $\frac{1}{8}$ inch long, with bright red eyes formed by some eight hundred image receptors. The front of its body is tan, and the back portion is black (darker in boys). Black rings run around its abdomen. Girls are about $\frac{1}{10}$ inch longer than boys.

Houseflies: about double the size of the fruit fly at $\frac{1}{4}$ inch long. Girls are still bigger than boys with a much wider space between their eyes. (Of course, if they're that close, you're probably not noticing the eye gap.) They're gray, with luscious green eyes and four thin, black stripes across the thorax.

REPRODUCTION

After a romantic fruit fly date of sewage al fresco and then buzzing slowly around the fruit bowl, mating is a rather aggressive affair. After wooing his love-beast with a courtship song made by vibrating his wings,

Tsetse fly

he positions a pair of claspers to grab on to her abdomen with gusto. He thrashes around for about ten minutes (sometimes in midflight) before she kicks him away with her hind leg, and he jets off for another conquest. Barry White music is not required, and no one has to meet anyone's parents.

Afterward, females will deposit some four hundred eggs in fruit, meat, or any decomposing food they can find. They'll lay some two thousand eggs in a lifetime. And, depending on the temperature in your house, the eggs will hatch in about two weeks.

Scew worm larva

LIFE EXPECTANCY

Fruit fly maggots or larvae will molt several times and live for about a month. Houseflies live for fifteen to thirty days.

Housefly

DIET

The fruit fly's scientific name, *Drosophila melanogaster*, is Greek for "black-bellied dew-lover," and *dew* (and its homophone) is its desert island meal.

Fruit flies eat virtually anything sweet and decaying, and use enzymes to break down vinegars, sugars, and toxic levels of alcohol (stronger than table wine) into yummy bits they like.

Houseflies feast on decaying liquid. If you're serving it in your kitchen tonight, this fly's in. If it's solid, the housefly can daub it with a bit of spit or barf, and poof—liquid dinner!

FYI: "fly specks" on walls are usually fly poop or regurgitated food.

PARADOX

Fruit flies drive a lot of us to distraction, but they pollenate flowers just as bees and wasps do, and they kill some genuinely harmful insects. And these fast-reproducing and fast-maturing critters have genetic gifts that can set fire to the most jaded biologists. They're probably the most commonly studied animal in the fields of genetics, physiol- ogy, and evolution, and even easier than sea monkeys to care for as pets. So, not so fast there with that bug swatter, tough guy!

FOR QUEEN FANS ONLY

Ancient Hebrews insulted their rivals by trashing their worship of Beelzebub (hero of the song "Bohemian Rhapsody," Queen, 1975), or "Lord of the Flies." Since flies live in poop, this was considered a mighty diss indeed.

And speaking of music, houseflies may be disgusting, but they are—surprisingly—musical. Their buzzing is made by beating their wings, and the faster they beat their wings, the higher the pitch of the sound produced. The housefly hums in the key of F, in the middle octave, and it beats its wings about 350 times per second—more than 20,000 beats a minute.

Black fly larva

IRONY DEPARTMENT

When cars replaced horses on the roads in the early twentieth century, they were hailed as a way to cut down pollution since they reduced the number of manure-loving flies.

ODDITIES

The treacherous human botfly (*Dermatobia hominis*) has been reported in some homes. First it crawls around under your skin, then it emerges as a larva that stands about an inch out of your body. Or how about the delightful coffin fly, which so loves humans that it doesn't even need them to be alive to find them entertaining: it waits in a coffin until we arrive!

Housefly

Another house-loving fly you might come across, the cluster fly, has such a nasty and messed-up life that you almost have to ask, "Does Darwin know about this?" In spring, females lay eggs in the soil that hatch a week later and bore into the bodies of passing earthworms, sometimes through their genitals. They eat the worm from inside, poke holes in it for ventilation, and then jump out of their hosts and back into the soil to pupate before emerging again to head over to your place for some dinner and a movie.

YIPES!

A classic (but controversial) 1954 study of a pair of fruit flies concluded that if the female produced one hundred eggs, and all survived, after twenty-five generations there would be about ten to the forty-first power of flies after six months. The scientists said that "if this many flies were packed tightly together, 1,000 to a cubic inch, they would form a ball extending nearly from the earth to the sun."

Flies are some of the few insects that can fly backward! Also, the sudden presence of flies in the house may mean a gas leak—so beware!

GET THEM OUTTA HERE!

Eliminate flies? Ha! That's a good one. People do use traps, flypaper, and poisons, but clicking your red shoes together and wishing they'd leave you alone is just as effective.

Black fly

Ants

**MONOMORIUM PHARAONIS,
TAPINOMA SESSILE, AND
SOLENOPSIS INVICTA**

What's most scary about ants is not their ferocity (most of them ain't fierce). It's their machine-like efficiency in spreading bacteria. They're the toughest of all household pests to control. They're too organized to fail and too cool to fake mean. They're the storm troopers of the insect world and they want you out of your kitchen—now!

With ants, all the action ends up happening in the kitchen—just like at a college mixer. They love your kitchen easily as much as you do. And as long as there's a steady supply of sweet, sticky foods in it, omnivorous ants will always hang around. Especially for melon. Melon has the texture they love, the softness and sugariness. It makes them giddy.

Pharaoh ants (*Monomorium pharaonis*) carry more than a dozen dangerous bacteria, including staphylococcus and salmonella. But they rarely bite and don't sting. Odorous house ants (*Tapinoma sessile*) get their name from the pungent, rotten coconut smell they give off after you crush them. They live for several years, and carry antennae full of ugliness to your food supply. Malevolent fire ants (*Solenopsis invicta*) are epidemic in the Southern United States and in places like Southern California, and their range is spreading. They attack kids, pets, and the bedridden. They killed an ailing sixty-six-year-old in Mississippi in 1998, stinging her hundreds of times. Just don't make them mad.

APPEARANCE

The reproductives all have four wings, distinctive bent antennae, and boast dainty, pinched waistlines.

Pharaoh ants (also sometimes called "sugar ants" or "piss ants"!) are among the smaller ants, sizing up at $1/12$ to $1/16$ inch long. Their bodies are light tan to red.

Odorous house ants are your Acme brand, black and brown jobs. They're $1/8$ inch long and—like all ants—have segmented, oval bodies.

Fire ants are (no points for guessing) red. They're also bigger than most ants, sometimes twice the size of odorous house ants.

SOCIAL LIFE

An ant's nest, part of a cozy colony of up to a million or so siblings, in-laws and hangers-on, holds one or more pampered queens and armies of sterile female worker ants to trawl for the queen's mates and food, and tend her babies. The workers roam freely in and out of your house foraging, leaving chemical trails behind them to alert relatives and pals to promising goodies. But if you want to kill the colony, you've got to follow the trail all the way to the nest. Killing the foragers is useless. Kind of mean, too, actually.

Black ant

REPRODUCTION

Queens lay eggs every day, and in about three weeks the eggs are adult workers.

TAKING FLIGHT

Yes, some ants fly too. The real nuisance in the house is that, at certain times of the year, colonies will produce a handful of reproductive and winged males and females, whose job it is to fly off and set up new colonies.

Usually, they'll set up that colony somewhere in or near your house. And it's one of these winged females, pregnant, who will shed her wings and become a new queen.

RANGE

They can live anywhere in your house, especially near warmth and water. And if a colony feels the vibe isn't right where they are, they're happy to pack up like the Joads and move the entire brood to a new spot—overnight. Most ants love light sockets, house plants, attics, floorboard cracks, and any other tight, cozy space.

GET THEM OUTTA HERE!

Ants are among the most difficult pests to control. In fact, some say spraying ant trails only makes the problem worse. The best method is to find the nest and spray it with insecticide. Trouble is, the nest is probably going to be invisible to you unless you strip your house down to its two-by-fours. You could, though, try to find a nest by leaving out some yummy food for them, watch them come for it, and then see where they go after they've chowed. But is it just us, or is this a little obsessive?

WELL, THIS MIGHT WORK

Insecticide fog sprays aren't very effective. They're like using slingshots against an infantry brigade—you can pick off a few stragglers, but you won't make much of a dent. A mixture of one part boric acid per one hundred parts ant food (bacon, melon, and the like) placed in a bottle cap and left strategically around the house (away from kids and pets) can kill a colony. Oh, but you have to wait about a month for the mass die-off.

DOES DARWIN KNOW ABOUT THIS?

It's hard to know exactly what the Great Maker had in store when he/she created fire ants. They're just vicious, tenacious bastards with a wicked temper, and will bite you over and over again just for the hell of it. They've been in the United States since arriving from Argentina in the 1930s, on boats docked in Mobile, Alabama. They now infest some five hundred square miles of Orange County, California.

At least eighty Americans have been killed by fire ants since their arrival here.

Ant larva

Cockroaches

BLATELLA GERMANICA AND PERIPLANETA AMERICANA

The ravenous and iconic cockroach is the world's favorite insect punching bag. We're as horrified to find them in our houses and apartments as we are delighted when we squish them underfoot. But before high-fiving yourself over a roach kill, ask yourself if you can do any of the following: go three months without food, survive hours without oxygen, laugh at a nuclear blast, and even live without your head for several days. Thought not.

The roach is an evolutionary masterpiece of engineering —from its bendable, waxy shell that allows it to burrow into tight cracks (the width of a quarter will do), ferocious jaws, ultrasensitive antennae (they can smell water), and steroid-pumped frame, it's the Maserati of the insect world. And it'll munch on you dead or alive. The planet's oldest insects are whip smart (they can figure out how to navigate a tricky maze after five tries) and view nuclear Armageddon as a speed bump.

APPEARANCE

The most common species are German cockroaches (in Germany, they're—not surprisingly—called "Russian cockroaches"). They're about 1½ inches long, and pale brown.

Less common are American cockroaches, which are much bigger at up to 2 inches long and a darker brown. Germans have wings but don't fly. Americans have wings but fly badly.

Even though both are toothless, they can rip, tear, and grind food—with malice.

Both species have antennae larger than their bodies, made up of one hundred to two hundred flexible segments for maximum mobility. And ultrasensitive hairs on their legs give them extrasensory input. Two thousand lens compound eyes let them see in almost all directions at once.

German cockroach

GENTLEMEN, START YOUR CERCI

This is pretty nifty. Roaches have ultrasensitive hairs near their rear ends called "cerci," which send lightning-fast messages to the brain. From the instant you walk into your apartment, even before you turn the light on, the cerci will shoot a warning message to the brain, and back again, with orders to flee—in $1/20$ of a second.

Wood cockroach

If you walk into your place and roaches don't run away or hide, you've got serious problems. That means you are so badly infested that the slackers have been crowded out of their hiding places.

DIET

Whaddya' got? These guys really put the *omni* back in *omnivorous*. They'll eat anything we'd eat, plus other roaches (dead or alive), people (ditto), poop (their own or anyone else's), glue, hair, chips of concrete. These guys have iron guts.

When they chew on living people (babies or the elderly, say), it's to get at food on a person's face— not the face itself. But we know this isn't very comforting.

For some reason, roaches seem to hate cucumbers. But they love warm beer. Oriental cockroaches, which you will occasionally see in the United States or Europe, love cinnamon buns above all.

RANGE

German roaches love kitchens and bathrooms more than any other rooms, but can get cozy anywhere in your house, especially in places like the gaps between tiles, behind stoves and refrigerators, and in any cracks in. wood or caulking.

The sulky Americans, on the other hand, are a little more shy at home. They're more likely to be found in or near basements, sewers, boiler rooms, or water heaters. Just think damp, dank, and dark.

American cockroach

In the wild, the 3,500 to 7,000 or so roach species are found everywhere—in snow and jungle, underground and the Alps, in rain forests and deserts.

REPRODUCTION

In general, German roaches are more reproductive than the Americans, but they both know how to fill up the nursery. A German female can produce up to 400 babies in her 150- to 200-day life. One oft-cited U.S. Food and Drug Administration study found that one fertilized female could —in theory—produce 10 billion females in just 1½ years. And the unfertilized eggs of American females can produce more females—without any sperm!

Cockroach sex happens year-round and is, as you might have guessed, a rather odd and aggressive affair. First, females give off pheromone scents that let the boys in the area know she's up for it. Then they both start producing high-pitched love songs to perfume the air with romance. If the boy is game, he'll lift his wings and invite her to mount him. She'll mount him, but, before the action gets started, she'll drink a special substance near his neck that he's produced for just this occasion. If she doesn't get this special ambrosia, she'll dismount. If it goes down easy, she'll turn around till they're rear-to-rear, and Tantric mating gets underway. This can last an hour.

LIFE EXPECTANCY
German roaches live 90 to 200 days, and American roaches average about 440 days.

GET THEM OUTTA HERE!
Americans spend several hundred millions of dollars per year trying to kill roaches, with a wide range of methods and very mixed results.

Some, for example, try live geckos, who love eating roaches. But be warned—geckos make very loud noises munching roacharitos.

Also used over the years to eat roaches: spiders, hedgehogs, and parasitic wasps.

You could "stress them to death"—by putting them in jars and rotating them until they become so stressed they die—or freeze them to death. As the temperature heads south of 15 or so, they'll die— if you don't first.

Roach egg cases

Rolled-up newspapers for one-off thrill kills work great, even though it doesn't tackle the real problem. But even with this trusty old method, be careful. Young roaches have at least fourteen "breaking points" from which they can part and still survive.

DAVID VS. GOLIATH

The challenge is that, since roaches are so prolific, every time they figure out how to survive a gel or bait, this resistance becomes embedded in their genetic code. This "memory" is then passed on to their descendants. So after a while, some of those gels, sprays, baits, and powders will lose their effectiveness Also, roaches learn fairly quickly, thanks to an amazing sense of smell, to stay away from places that have been sprayed.

Some insect populations, like grasshoppers or crickets, are kept down in places where humans eat them. But a roach's smell is so bad, and their diet is so repellent, that most cultures stay away from roaches.

Among the roaches' favorite places to hang out are food-rich supermarkets. And a common way to bring them home is in grocery bags. So check these before putting groceries away.

Some places
will try to sell
you sound-emitting
devices that are meant to
scare roaches. But if you listen just
above the din, you can hear the roaches laughing.
We don't recommend them.

KAFKAESQUE

In Franz Kafka's classic *Metamorphosis*, Gregor Samsa famously wakes up one morning to find himself turned into an insect. Most people—fascinatingly—assume he's a cockroach even though the book itself never mentions what kind of insect he becomes. The iconic roach, so root-ed in our psyches as the most vile of vermin, is morphed into Kafka's mutant.

THE THINGS THEY CARRY

Roach body parts, poop, and body fluids are thought to play a role in the rising rates of asthma, especially among kids in the inner city. This is the most widespread health problem they cause.

They also carry aspergillus (potentially dangerous mold that grows in house walls). In fact, about forty species of disease-causing bacteria have been found in roaches, including bubonic plague and leprosy. They can also transport polio and hookworm.

DONNER PARTY, TABLE FOR TWO

When a colony gets too crowded, roaches will resort to cannibalism to ease population pressure.

IRONY DEPARTMENT

Cockroaches are very careful about their personal hygiene. They frequently clean their antennae by locking them in their legs and running them through their mouths.

SLIPPING INTO SOMETHING MORE COMFORTABLE

German cockroaches molt about six times in their lives. It's the only way they grow. They dump their old exoskeleton by gulping air and squeezing it off. After they slip away from their old body, they gulp air again to create a new exoskeleton. New ones look albino for a short time, until they turn golden brown.

SATURDAY NIGHT'S ALL RIGHT...

... for antennae fighting. When two cockroaches in a rotten mood meet, they rumble with antennae drawn like swords, followed by threat postures like walking high as if on tiptoe.

Males fight one another, as do males and females. Fights are rarely to the death, but the loser may get his legs bitten off. Curiously, antennae-fighting is sometimes done as foreplay too.

CONGRESSMEN AT LARGE

In 1982, a team of crack entomologists was summoned to hunt down a fast-multiplying brood of roaches that had overtaken parts of the House of Representatives building in Washington, D.C. The bug SWAT team was stunned to find an incredibly resistant strain that to this day holds a special place in the hearts of some entomologists. It's known as the HRDC (House of Representatives, D.C.) strain.

ONE GIANT LEAP FOR ROACH KIND

There was reportedly a roach stowaway on *Apollo XII,* a craft that eventually landed a vehicle on the moon and could have disgorged a roach or more on the lunar surface. A small colony of roaches was sent into space in 1998, and some have reportedly been spotted on Russia's MIR orbiting space station.

COME UP AND SEE ME SOMETIME

Cockroaches have their own hall of fame in Plano, Texas. The museum features roaches dressed as historical figures and celebrities like "Liberoache" and "David Letteroach." And don't miss the Bates Roach Motel. "La Cucaracha," sung all the way back to the days of Pancho Villa's troops, is the roaches' (but not the Roches') most famous musical legacy. And movies like *Men in Black,* MTV's *Joe's Apartment,* and the bomb *Vampire's Kiss* (where Nicolas Cage eats a roach) have enshrined them on the silver screen.

ODDITY

Roaches are so easily adaptable that there's a roach for virtually every microenvironment on Earth. One of our favorites is known as the brown-banded or "TV roach" (*Supella longipalpa*). They're small at just a ½ inch long, and they love TV cabinets and other warm, dry places like alarm clocks and stove clocks. They're legendary reproducers, giving birth in German roach–caliber numbers.

THE FLIP SIDE

Roaches aren't all bad. In some places, they're thought to benefit us. In some cultures, ground-up cockroaches mixed with sugar are applied to ulcers and cancers as a salve, and during the slavery period, American slaves made cockroach tea for tetanus, and spread boiled cockroaches over wounds.

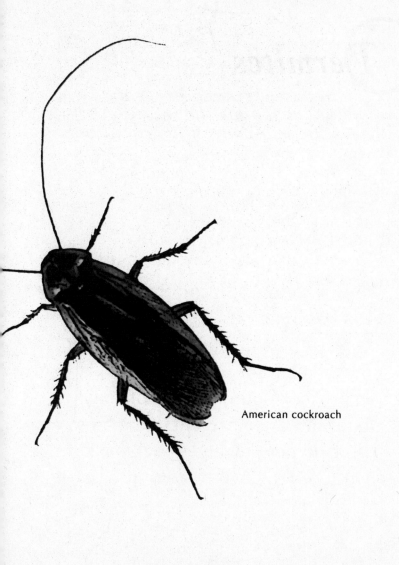

American cockroach

Termites

RETICULITERMES HESPERUS,
INCISITERMES MINOR, AND
ZOOTERMOPSIS ANGUSTICOLLIS

Termites love wood. Can't get enough of the stuff. They'll also chow on books, furniture, rifles, and baseball bats. In a pinch, they can toss back leather, cloth, rubber, and plants. But wood is their ambrosia. A puny colony of sixty thousand or so can eat five grams of wood per day and knock back a two-foot-long two-by-four in a year.

They will eat your house to the foundation if given enough time—studs, joists, and all. And they embed themselves so deeply in the "bones" of your house, that you usually only find them after it's too late.

Oddly, most termites can't digest wood on their own. They need microbes in their gut to produce an enzyme that breaks down the cellulose in wood. Parents then pass on these useful microbes to their kids by "anal feeding" (yup, just like it sounds).

Colonies can easily grow to several million strong and, just like ants, form amazingly organized societies.

They are divided into castes: workers, soldiers, and reproductives. The lucky queen and king are the randy, royal reproductives. They lead hedonistic

Soldier termite

lives that would have made Barry White blush. These tantric termites lounge and mate continually, hanging out all day and night in cavelike love nests in case the mood strikes. The queen sometimes lays three thousand eggs a day.

The cream-colored workers are at the bottom of the pecking order. For a worker termite, days are hectic rounds of finding food, raising the queen's babies, and digging and cleaning tunnels. But with no wings, there's no hope of ever leaping off the treadmill. Outside colony walls, huge-jawed soldiers wage battles to the death with predators like ants, which snack on termites like M&M's. But until we start popping them like candy, termites will continue to dodge the many millions we spend each year trying to kill them.

Soldier termite

APPEARANCE

The reproductive royals are the largest termites, followed by the soldiers, and then by the most abundant caste, the workers. Reproductives are dark brown with brown wings, and about an inch long. They're white, tan, or black in color. When pregnant, which is to say most of the time, the queen swells enormously and dwarfs her king in size.

Soldiers have big, orange rectangular heads and powerful jaws, and spend their time at the colony's openings squirting a warm liquid latex out of their big heads at intruders (ants, mostly). One species that doesn't have soldiers to defend the colony will spill all its gut contents on invaders, martyring themselves in the process.

The blind, soft-bodied workers are about $\frac{1}{8}$ inch long, depending on the species. They, like soldiers, are sterile. Workers spend their days working, and—perhaps not surprisingly—live just two years.
Among their fun tasks?
Barfing or pooping
up food for the
soldiers.

Termite queens

Winged termites are very similar in appearance to some winged ants, and are sometimes even called "white ants." Some key differences—termite antennae are straight (not elbowed), termite waists are plumper, and ant wing sizes differ (while all four termite wings are the same size).

REPRODUCTION AND SWARMS

Reproductive kings and queens (also called "swarmers") as young adults have wings, which they use to fly off to set up a new colony, usually in late summer or fall. Once in a new locale, a couple mates, sheds their wings, and finds a good nesting site to build a small, private nursery. They seal it off from the outside

Worker termite

and the
queen lays
thousands of
eggs. After the eggs
hatch, the larvae skedaddle.
After three to four years, the colony
will produce new winged swarmers,
which will fly off by the thousands
through tiny tunnel holes to start the
whole cycle again.

Soldier termite

But flying off to start a new colony is a very hazardous affair, and only about one in one thousand swarmers will become new colonizers. Their wings are so bad and their flight is so shaky that all their enemies—ants, spiders, lizards, rats, birds, and even humans—will have an easy swipe at them midflight.

LIFE EXPECTANCY

The queen and king's luxurious lifestyle, and their lifelong fidelity to each other, must keep them fit, because they can live for twenty-five years. The chaste soldiers only live for a few years.

DIET

Fir *a l'orange,* pine *con pinos,* redwood *avec* artichoke hearts, spruce scallopini. It's all wood to them.

Dampwood termites like their wood moist and are happiest in coastal mountain or humid areas. Subterraneans like it moist too, but—as

Drone termite

you'd guess—tend to find their dinner, mostly rotted wood, below ground. Drywood termites like their wood dead but not yet decayed, and love to feast on building lumber, utility poles, and the like.

RANGE

Underground termite tunnels can go for one hundred yards and more, poking up at wooden wet bars all along the way.

GET THEM OUTTA HERE!

Make sure all building wood is at least twelve inches above any soil, keep all stucco siding off the ground (they love the stuff), and keep attics and foundations dry.

One of your best weapons is a sand barrier laid all around the house foundation. Termites can't build their tunnels in sand, so they can't get to the wood underneath.

Use termite-resistant woods in new buildings. Douglas fir is moderately resistant, while hemlock and spruce are termite magnets. Some termite-friendly woods can be made repellant if treated with chemicals or pressure-treated.

For drywood termites, there's a huge arsenal at your disposal. Try killing them with chemicals, heat, freezing, fumigants, microwaves, electrocution. Hell, try a firing squad. Fumigants using effective poison can kill a whole colony in three days. Trouble is, they may do the same to the foods and people left in it. Likewise for heat-killing, which involves heating your whole house to 120 degrees. This will kill all the termites, but what will it do to your house? Microwaves kill by boiling the fluids inside termites' cells, and ninety-thousand-volt termite tickles drilled directly into the wood are also deadly.

For subterranean and dampwood termites, though, only insecticides or baits will do, since they live underground.

THE ENEMY OF MY ENEMY IS MY FRIEND

You can also use what the trade calls "biological control agents," a fancy term for sicking some other insect baddies on them. Up for this entomological roulette? Try fire ants or nematodes.

TERMITE FARTS

According to a 1982 article in *Science* magazine, termite farts were responsible for up to 30 percent of the Earth's atmospheric methane, though several other studies since then have said the figure is lower. And we thought it was mostly from cows, no?

THE GOVERNOR JUST CALLED: YOU'RE FREE

Not all termites are useless house eaters in need of a good whoopin'. OK, maybe most are. But some are to be celebrated. They're needed to recycle plant and wood material, and their tunneling in the ground helps keep soil loose. On the other hand, they're probably snacking on your deck as we speak, so if you still feel the need to go Rambo on 'em, we'll forgive you.

COME OUT, COME OUT, WHEREVER YOU ARE

Termites are pernicious because they're silent and almost always invisible. One of the most popular ways to find out whether and where you might have an infestation is to look for so-called kick-out holes, BB-size holes in their tunnels through which they push out their poops.

Soldier termite

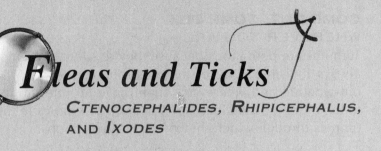

Fleas and Ticks

CTENOCEPHALIDES, RHIPICEPHALUS, AND IXODES

Tiny fleas and ticks have had a big, bad rap for a long time. One of Shakespeare's characters in *Henry V* refers to a flea as a "black soul burning in hell's fire." The Jewish Talmud allows such vermin to be killed (smote?)—just not on the Sabbath. And no less a judge of character than Aristotle himself called ticks "disgusting parasitic animals." Others are not so polite.

Fleas and ticks are long overdue for an image overhaul, but as long as they (fleas) leap 150 times their body length through the air for a swig of our blood and shove toxin-filled noses into our capillaries (ticks and fleas), they'll be reviled. These micromonsters are not only expert at puncturing human skin and sucking blood, they're just as good at clinging on to us and making it nearly impossible for us to kill them.

Sheep tick

Furry pets like cats, dogs, or rabbits are the chariots on which fleas and ticks ride into our homes. But once inside, they'll gladly lunge cross-species for the 24/7 Bloody Mary (or Michael) cocktail lounge that is *corpus humanus*. They'll happily take their disease-spreading, parental-poop-eating, double-penis-wielding, mating-while-eating freak show *chez vous*. And we shouldn't be surprised. After all, we let 'em in.

Cat flea

APPEARANCE

Fleas up close look a bit like lobsters, with overlapping scales running the length of their bodies. But they are flat and wafer thin—the better to navigate undetected in and around human and animal hairs and fur. And most fleas have a smooth and firm shell, making them harder to yank off our pets or ourselves. These wingless wonders range from 1.5 to 4 millimeters long.

Human flea

But fleas are most famous for their amazing knife-and-straw combination mouthparts, and six superhero legs (more on these later).

Deer tick

Ticks also boast spectacular skin-busting and bloodsucking harpoons for mouthparts, but in a rounder and tidier package than the flea. Most ticks are less than one millimeter long unfed, but swell up like balloons after a hefty blood meal. Unlike most bugs, they have no segments in their bodies. They're kind of shapeless blobs with colorful, complex patterns. And girls are both bigger and prettier than boys—a gross breach of beastly protocol.

There are several thousands of species of both fleas and ticks, and each is associated with a specific animal, i.e., dogs, cats, rabbits, beavers, deer, and the like. But most ticks can (and will) feast on an array of mammal hosts, including humans, not just the species that's got a real jones for us specifically (*Pulex irritans*, the human flea).

JUMP THIS!

The flea's astonishing jumping ability relies on a compound that would make any Tour de France cycling coach drool. Who wouldn't treasure an athlete who could leap up to 150 times his height and horizontally by 80 times? A flea's long and gangly legs contain resilin, an elastic protein that stores energy and releases it in bursts of g-force that could easily launch an Apollo rocket into space. Stop-motion photography reveals the delicious fact that fleas are absolutely and utterly out of control when they tumble through the air. They don't seem to care what part of their body they land on. They're so well-armored that it doesn't matter.

If it's any consolation, fleas can't fly. At least not yet.

Dog flea

MEALTIME

Both fleas and ticks have ingeniously devised tools for ripping into our skin and depleting our precious bodily fluids. And they may suck for hours to get just what they need. Creepily, to find their hosts, they don't even need to see them—they can just detect their body heat and carbon dioxide emissions. But don't worry too much—all they want is about 0.0004 of a cubic millimeter of the hot, red stuff. And they're patient: an adult can go for six months hunting for the perfect bloodbath.

Mouthparts of a tick

Fleas boast long, sharp beaks for poking and sucking. Before biting, a flea will raise its tail high in the air, and then shove its proboscis down hard, tenderizing the kill zone. Then, it will lubricate the area with saliva that contains an anticoagulant and local anethestic to prevent clotting. To cling on ever so tightly, it will move into position a set of backward-facing comblike teeth that give extra grip. Then, let the Games begin.

Ticks have amazing mouths that boast a harpoonlike dagger that would have made Bluebeard blush. It's flat on top and curved on the bottom, covered with an awesome array of curvy barbs for a tenacious grip. The harpoon (called a "hypostome") is both a knife and a straw. As a backup in case the harpoon doesn't do the job, many ticks also secrete a cementlike gooey fluid around the sucking site for extra stickage. After cutting blood vessels under the skin, blood pools in the area and the tick simply slurps it all up! Like the flea, a saliva spread around the area contains an anticoagulant. But it's this very saliva that contains an arsenal of toxins that can hurt or even kill humans.

SLOW FOOD

You will not enjoy hearing that tick meals may last several days. Unless you catch and kill them first (fat chance), a tick will

Cat flea

spend languorous hours picking, poking, suck- ing, and digesting you and your blood for the better part of a week

LOOSEN THAT BELT!

To accommodate their ravenous eating binges, ticks have an accordionlike stretchy underbelly that can expand its

Sheep tick

stomach by some twenty-five to fifty times to hold down all the blood. Talk about "relaxed fit"!

MATING

Flea mating is a multihour Caligulaesque bacchanal that, in fine flea fashion, always starts with a big meal. But, in a kinky twist, mating occurs all the way through the meal, with both sides coupling and quaffing blood at the same time. The male needs the blood for stamina (and who can blame him?), while the female needs it to ovulate. The flea penis is an anatomical wonder, comprising two different shafts wrapped around each other. One is for penetration and the other transmits sperm to the female. Both parts emerge from the flea's rear end.

OH, THAT'S NASTY!
Female fleas lay their eggs right on top of their hosts, a few hours after mating. After coitus, she and her mate will eat and then defecate profusely, over and over, leaving "food" for their creamy white wormlike larvae when they hatch.

LIFE CYCLE
Flea moms and dads also give their precious newborns, who arrive some two to twelve days after mating, even more snacks—including predigested blood chips that they produce just for this purpose. These hors d'oeuvres complement

Ticks eating

the larvae's staples of skin scales and their host's loose hairs. Some really clever parents even try to synchronize their litters with those of their hosts, such as rabbits, so babies are assured of a steady blood supply right out of the gate.

Ticks have a life cycle of three stages (larva, nymph, and adult), and each also requires a blood meal to take the next step. Disturbingly, even though they have a high mortality rate, they also have long lives—up to three years (yipes!). One lab tick even survived four years—with its head cut off! Females lay up to six thousand eggs at a squat. Not surprisingly, they drop dead right afterward.

THE THINGS THEY CARRY

Fleas carry numerous diseases that plague humans.

These include typhus and bubonic plague. Fleas riding on rats' backs killed one out of three Europeans during the Middle Ages, forever ensuring themselves a place in the beastly hall of fame.

Ticks bring us such maladies as Rocky Mountain spotted fever, tularemia (a bacteria from rabbit ticks that can cause ulcers and pneumonia-like symptoms), and the famed Lyme disease. Lyme is spread by deer ticks, and may leave a ringlike red rash on the skin a few days to weeks after a bite. Left unchecked, fever, stiff joints, and chills follow, and later meningitis, paralysis, and abnormal heartbeat. It can be fatal.

GET THEM OUTTA HERE!

Flea bites can be maddening. Fleas usually do their best work around the ankles and legs, leaving bite marks in clusters of two to three. And there's no such thing as a digestive stroll after dining. These guys will work a rich vein of blood over and over again, for hours on end. So you may mistakenly feel you or your house is "infested" with them when it's often just a few very busy critters.

Sobering thought: the extreme measure of getting rid of your pets may not make a dent in your flea problem. Larval fleas spend all their time off their hosts, and even adults roam elsewhere some 90 percent of their lives. Only a full-front chemical assault with foggers and chemical "bombs" on your carpet stands a chance.

Ticks are a nightmare to try to get off. If you yank their head off, their mouthparts and body will keep on slurping your blood and can still cause infection. You could try burning them off, with all the attendant risks. Or you could try dousing it with gasoline or alcohol. Just don't try both methods!

As a tick precaution, wear long-sleeved shirts and pants, and tuck everything in before you head out into the woods. Light-colored clothing allows for easier post-hike tick ID. And drench your body and clothes with repellent.

Cat flea

Fabric Fiends and Pantry Pests

ATTAGENUS UNICOLOR AND
XESTOBIUM RUFOVILLOSUM

Herein, a salute to the unsung denizens of the domestic jungle, a grab bag of lesser bugs like moths and beetles, which can wreak terrible havoc in the kitchen, closets, and floors—and barely leave a trace.

The deliciously named death watch beetle and cigarette beetle are some members of this shadowy clique. The team also includes flour moths and drugstore beetles (*Stegobium paniceum*), and one of our favorites—the confused flour beetle (*Tribolium confusum*), so named because it's so confusingly similar in appearance to the red flour beetle.

The death watch beetle is usually only detected after it's devoured the furniture it's feasting on; the drugstore beetle (cousin of the tobacco-loving cigarette beetle) will gobble up medicine and spices, or anything not bolted to the bathroom mirror, and even aluminum; and the Mediterranean flour moth will indiscriminately feast on most anything in the pantry. Dog and cat food is their catnip. These pests live unusually long lives, can survive without water, and even feast on some of our "insecticides."

Confused flour beetle

They're a colorful, handsome, and devious bunch on the whole, and their mellow demeanor fools us nightly. They are like an incubus that unsettles our sleep, quietly gorging on the cereals we plan to feast on the next morning, and even laying waste to the table we plan to eat it on.

MEET THE BEETLES!

Beetle larvae are legless or have just three pairs of stumpy legs, all near the head. Moth larvae have three pairs of legs near the head and more leglike appendages near the abdomen. With beetles, both adults and larvae feed and do damage, while with moths it's usually only the larvae that do so. In fact, some adult moths even shed their digestive systems altogether.

Carpet beetle larvae

Most of the pantry beetles (flour, drugstore, cigarette, etc.) are about ⅛ inch long, cylindrical, and light brown, with a signature rounded body shape and hard wing covers.

Indian meal moths have red brown wings, with a coppery sheen, and a white gray body. They're often confused with clothes moths, but clothes moths are smaller and hairier. Creepily, some shades of Indian meal moths' coats are camouflaged with the muted brown white color scheme of many popular kitchen appliances.

HI, MR. DEATH WATCH!
WELCOME TO WOOD-EATERS ANONYMOUS!

The story of how the death watch beetle got its name is almost as juicy as the name itself. It seems that a female, when she's in the mood for love, will slap her head against whatever woody treat she happens to be eating. It's an extremely faint sound, but if enough of them do it at once, it can become audible to humans and brings on a fearful, haunted feeling that is compared to a "death watch." Edgar Allan Poe referenced one in "The Tell-Tale Heart." We can't make this stuff up.

THE BEETLE AND
THE DAMAGE DONE

Confused flour beetles, like many men late at night in their BVDs, just love eating heaps of cereal. In fact, anything in the pantry will do, including crackers, peas, spices, dried noodles, and anything with flour.

Death watch beetle

Carpet beetles love all
rugs and natural fab-
rics. They'll also
gladly eat fur,
leather, and
wool, and
can do even
more damage
than clothes
moths. They're
the most difficult
of all fabric pests to control,
and will eat anything from dead
animals to piano felts and violin bows.

Powder post beetles, just $1/16$ to $1/8$ inch long,
leave pinhole-size gouges in wood, and a
fine powdery dusting underneath.

Carpet beetle

Clothes moths love, above all, dirty clothes, especially
those with sweat, food, and drink stains on them. They also
prefer clothes with animal proteins in them, like leather,
fur, and the like. They're not wild about cottons or linens
or synthetics like rayon. Their true genius is that they can
digest the seemingly undigestible substance called kera-
tin, found in animal proteins and hair-derived tissue.

DAILY DIARY:
DEATH WATCH BEETLE

9:00 p.m. – Woke up, went looking
for wood

9:10 p.m. – Found some wood,
started eating

1:30 a.m. – Mated, then back to
the wood

7:15 a.m. – Humans arrive, laid
low

8:30 a.m. – Humans gone, hit
the wood

9:00 p.m. – See: Yesterday

Death watch beetles are the
24/7 eat-and-mate type, tunneling all day in wood, and
turning it to sawdust. Unlike termites, they can even eat furniture
that is not resting on the ground. In
fact, their wood orgies put some termite
species to shame. When they're not eating
wood, they're making baby death watch beetles
that will quickly grow up to be just like mom and dad.

Confused
flour beetle

MATING

Male and female Mediterranean flour moths have wildly
different ways of finding each other. While girls lift up
glands at the tip of their abdomens, and let pheromones
waft across to their intended, boys lift up their glands and
slap would-be lovers upside the head and thorax.

LIFE CYCLE

After mating, which can sometimes end with either partner being eaten by the other, flour beetle (red flour and confused flour) females will lay their eggs in the pantry. Eggs hatch in about a week to creamy brown white grubs that we sometimes call "bran bugs." Females can lay up to four hundred eggs in their lifetimes, which they will occasionally eat themselves. These beetles can live for a year, while some like the drugstore beetle may live just two to seven months.

SLASH AND BURN

Mediterranean flour moths and Indian meal moths will eat fruits, nuts, flour, cereal, corn, and grain. They'll also spin sticky cocoons around food and packaging, creating even more of a nuisance for us to clean up.

Also, many of them secrete really nasty-smelling chemicals called quinines while they eat. It's thought to be a way of marking territory, but it also spoils our food even more quickly.

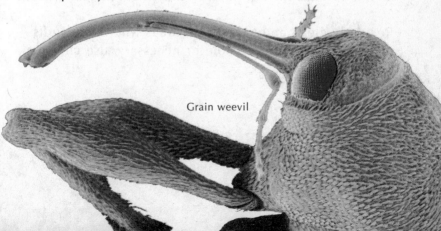

Grain weevil

GET THEM OUTTA HERE!

Food-eating beetles and moths are tough to kill off, especially since so many of them can live for weeks and months without food. Some guides advise storing all grains, cereals, and the like in the refrigerator or freezer, or heating them up to 175 degrees for ten minutes. If that's not practical, store such things in airtight containers. And, if you think your cedar closet will keep clothes eaters away, think again—cedar will lose its bug-repellent qualities over time.

Much more fun than freezing food are glue traps that lure male bugs using pheromone smells that scream, "You're gorgeous and I want you now!" Or tap into your inner sadist by sprinkling diatomaceous earth where beetles and moths feed. This is dirt with razor-sharp particules in it that will Cuisinart bugs, but is too fine to affect humans.

Some are also lured by sticky fish oils left out for them—if you can bear the smell yourself. Or unleash a knot of toads in your pantry—they'll eat them.

Regularly clean and vacuum all areas where foodstuffs are kept and, when possible or necessary, use chemical sprays like pyrethrum.

DEATH BY MANILOW

For wood-eating beetles like the death watch, surface spraying may be effective, but try not to kill off any spiders who are its natural predators. Pressure injection or irrigation by chemicals is another option, but it can eat away paint and plaster and create a fire hazard. Water injection could work, but causes wood to swell. All of it is a crap shoot. Some of our fans have even had luck blasting Barry Manilow's "Mandy" at a volume of 11 right near them. It's probably worth a shot.

Grain weevil

SOURCES

BOOKS

BIOLOGY OF BLOOD-SUCKING INSECTS
M. J. Lehane
HarperCollins (NYC), 1991

BUG BUSTERS
Bernice Lifton
Avery Publishing Group, Inc.
(Garden City Park, NY), 1991

BUGS OF THE WORLD
George C. McGavin
Facts on File (NYC), 1993

BUZZWORDS
May Berenbaum
Joseph Henry Press
(Washington, D.C.), 2000

THE COCKROACH PAPERS
Ricard Schweid
Four Walls Eight Windows (NYC),
1999

THE COMPLEAT COCKROACH
David George Gordon
Ten Speed Press (Berkeley), 1996

THE ENCYCLOPEDIA OF INSECTS
Edited by Christopher O'Toole
Facts on File (NYC), 1986

ENCYCLOPEDIA OF INSECTS AND
SPIDERS
Rod and Ken Preston-Mafham
Thunder Bay Press (San Diego),
2005

ENTOMOLOGY & PEST MANAGEMENT
Larry Pedigo
Culinary and Hospitality Industry
Publications Services
(Weimar, TX), 2006

A FIELD GUIDE TO GERMS INSECT FACT
AND FOLKLORE
Lucy W. Clausen
Colliers Books (NYC), 1962

INSECT LIFE
Michael Tweedie
Collins (London), 1977

INSECTS
Bob Gibbons
Harper Collins (NYC), 1999

INSECTS
Geoge C. McGavin
Dorling Kindersley (NYC), 2000

INSECTS OF THE WORLD
Anthony Wootton
Facts on File (NYC), 1984

INSECTS THROUGH THE SEASONS
Gilbert Waldbauer
Harvard University Press
(Cambridge, MA), 1996

THE LIFE OF THE FLY
J. Henri Fabre
Dodd, Mead & Company (NYC),
1913

LIFE ON A LITTLE-KNOWN PLANET
Howard Ensign Evans
E. P. Dutton & Co. (NYC), 1986

THE NATURAL HISTORY OF FLIES
Harold Oldroyd
W. W. Norton & Company Inc.
(NYC), 1964

NINETY-NINE GNATS, NITS, AND NIBBLERS
May Berenbaum
Univ. Illinois Press (Urbana), 1989

NINETY-NINE MORE MAGGOTS,
MITES AND MUNCHERS
May Berenbaum
Univ. Illinois Press (Urbana), 1993

PEST CONTROL FOR HOME AND GARDEN
Michael Hansen
Consumer Reports Books
(Yonkers, NY), 1993

THE PRACTICAL ENTOMOLOGIST
Rick Imes
Simon & Schuster/Fireside (NYC),
1992

THE SECRET LIFE OF GERMS
By Philip M. Tierno Jr., PhD
Pocket Books (NYC), 2001

SIX-LEGGED SEX
James K. Wangberg
Fulcrum Publishing
(Golden, CO), 2001

SOCIAL LIFE AMONG THE INSECTS
William Morton Wheeler
Harcourt, Brace & Co. (NYC), 1923

THE STRANGE LIVES OF FAMILIAR INSECTS
By Edwin Way Teale
Dodd, Mead & Co. (NYC), 1962

A TEXTBOOK OF ENTOMOLOGY
Herbert H. Ross
John Wiley & Sons (Sydney), 1948

TINY GAME HUNTING
By Hilary Dole Klein &
Adrian M. Wenner
Bantam Books (NYC), 1991

URBAN ENTOMOLOGY
William H. Robinson
Chapman and Hall (London), 1996

WHAT'S BUGGING YOU?
By Michael Bohdan
Santa Monica Press, LLC, 1998

WHAT GOOD ARE BUGS
Gilbert Waldbauer
Harvard University Press, 2003

OTHER

BugInfo.com

Harvard University School of Public Health

Iowa State Department of Entomology

North Carolina State University, Department of Entomology

Penn State University, College of Agricultural Science

PestProducts.com

Purdue University Cooperative Extension Service

PHOTO CREDITS

The stunning images inside the pages of this book were captured by a group of talented photographers working throughout the world. The work of these and other individuals sparked our curiosity and sense of wonder, and, in many ways, inspired us to write this book.

We would additionally like to acknowledge Photo Researchers, Inc., and Jackie Tolley in particular, who provided many of the images in the book, and assisted with image research and acquisition.

Photo credits by page:

Frontispiece: Marcelo de Campos Pereira

3/4/5: Louis De Vos

6/7: Joshua Abarbanel

8/9: Louis De Vos

10/11: Joshua Abarbanel

13: Dennis Kunkel

14/15/16/17: Dennis Kunkel

18: Photo Researchers, Inc.

19: Dennis Kunkel

20/21: Photo Researchers, Inc.

23: Dennis Kunkel

24/25: Composited image. Andrew Syred / Original image: Photo Researchers, Inc.

26/27: Composited image. Andrew Syred / Original image: Photo Researchers, Inc.

29: Andrew Syred / Photo Researchers, Inc.

30: Andrew Syred / Photo Researchers, Inc.

31: Louis De Vos

32/33: Composited image. Eye of Science / Original image: Photo Researchers, Inc.

34/35: Eye of Science / Photo Researchers, Inc.

37: Nolie Schneider

38/39: Steve Gschmeissner / Photo Researchers, Inc.

40/41: Gary Meszaros / Photo Researchers, Inc.

42/43: Eye of Science / Photo Researchers, Inc.

44/45: Steve Gschmeissner / Photo Researchers, Inc.

47: Ricardo A. Palonsky

48: Louis De Vos

49: United States Department of Agriculture

50/51: Ricardo A. Palonsky

52/53: Louis De Vos

54/55: Dennis Kunkel

57: Photo Researchers, Inc.

58/59/60/61: Louis De Vos

62/63: Susumu Nishinaga / Photo Researcers, Inc.

64/65: Louis De Vos

67: Photo Researchers, Inc.

68: Clemson University–USDA

69: Hannah Mason

70/71: Charles Schurtz Lewallen

72/73: Composited image. Original image: Hannah Mason

74/75: Clemson University–USDA

76/77/79: Photo Researchers Inc.

81/82: Louis De Vos

83: Erik Mielke

84: Dennis Kunkel

85/86/87: Louis De Vos

89: Dennis Kunkel

91: Darwin Dale / Photo Researchers, Inc.

92/93: Steve Gschmeissner / Photo Researchers, Inc.

93: Darwin Dale / Photo Researchers, Inc.

94: Steve Gschmeissner / Photo Researchers, Inc.

95: Louis De Vos

96: Dennis Kunkel

97: Photo Researchers, Inc.

98/99: Composited image. Volker Steger / Original image: Photo Researchers, Inc.

101: Steve Gschmeissner / Photo Researchers, Inc.

102/103: Public Domain

104/105: Eye of Science / Photo Researchers, Inc.

106: USDA Forest Service Archives

107/108/109: Dennis Kunkel

110/111: Biophoto Associates / Photo Researchers, Inc.

116: Photo Researchers, Inc.

Image of magnifying glass on the title page and chapter openers courtesy of Joshua Abarbanel.

Image of calipers on the title page and chapter openers courtesy of Post Mortem Studio Rentals—Major Medical Props.